Cal has

Cal has ✂.

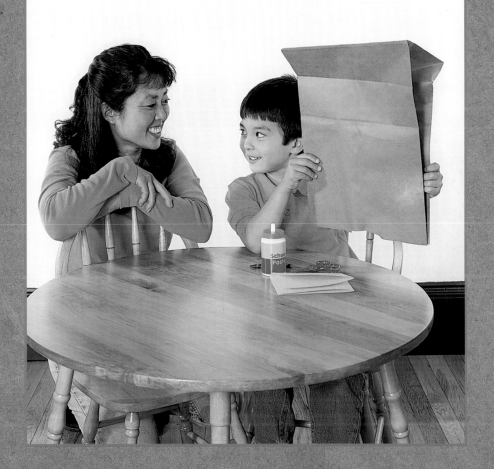

Hal has a big bag.

Hal has 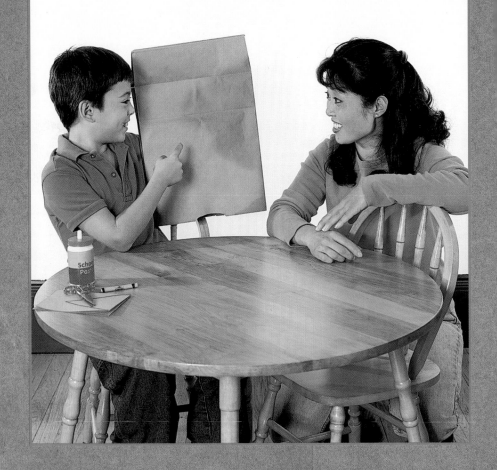.

Cal can cut it.

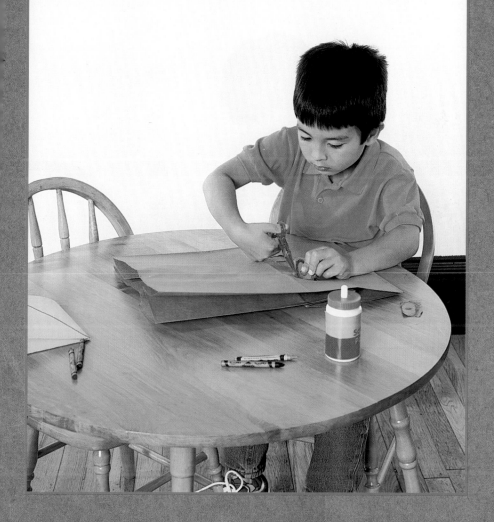

Hal can cut it.

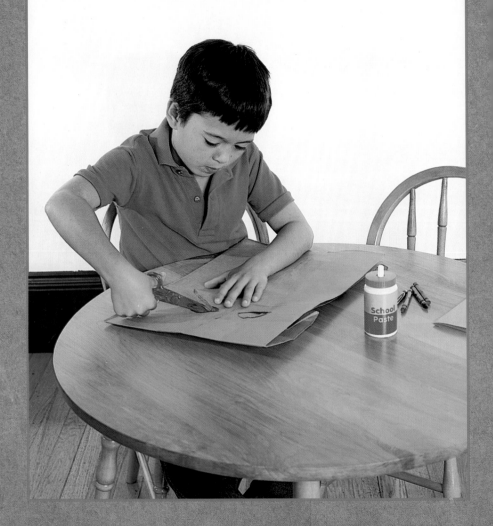

Cal got into the mask.

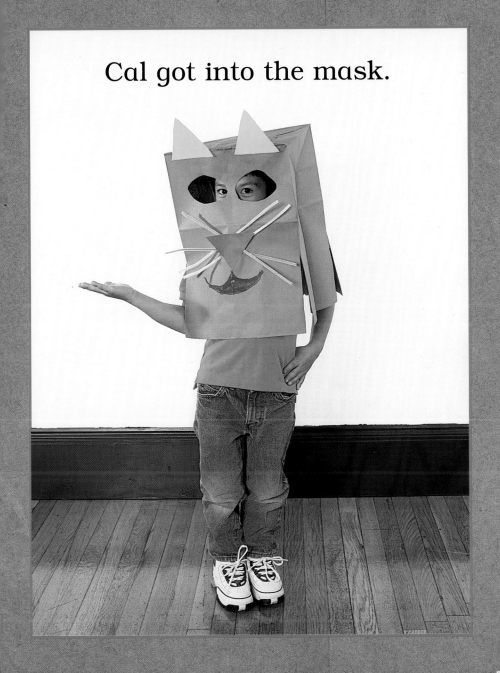

Hal got into the mask.

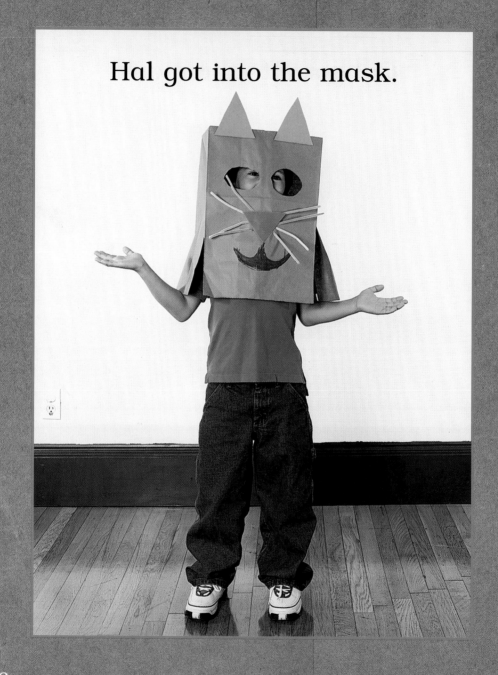

Cal asks, "Am I a cat?"
Hal asks, "Am I a cat?"

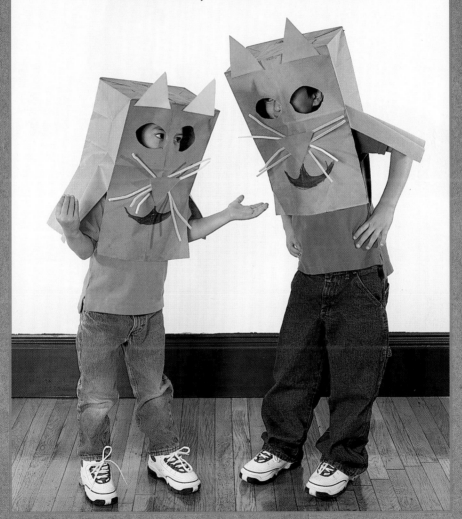

The cats can act!